Science Investigations

SOUND: AN INVESTIGATION

JACK CHALLONER

New York

Published in 2008 by The Rosen Publishing Group, Inc.
29 East 21st Street, New York, NY 10010

First Edition

The publishers would like to thank the following for permission to reproduce these photographs:
Alamy: 27 (Elvele Images); Corbis: 7 (Bryn Colton/Assignments Photographers), 21 (Annebicque Bernard/Sygma), 24 (Julio Donoso/Sygma), Cover and 26 (Ted Spiegel); NASA Images: 16; OSF/Photolibrary: 4 (Walter Bibikow), 5 (Harmon Maurice), 8 (Gerard Soury), 11 (Claude Steelman), 13 (Ingo Arndt), 19 (Mauritius GMBH), 20 (Tony Ruta), 25; Redfern: 18 (Rosie Hartnoll), 28 (Nicky Sims), 29 (George Chin); REX Features: 14; Science Photo Library: 12; Topfoto: 6, 10 (Joe Carini/The Image Works), 15 (Ellen B. Senisi/The Image Works), 17 (Jeff Greenberg/The Image Works), 22 (UPPA), 23 (The Arena PAL Picture Library).

Editors: Sarah Doughty and Rachel Minay
Series design: Derek Lee
Book design: Malcolm Walker
Illustrator: Peter Bull
Text consultant: Dr. Mike Goldsmith

Library of Congress Cataloging-in-Publication Data

Challoner, Jack.
Sound : an investigation / Jack Challoner. — 1st ed.
p. cm. — (Science investigations)
Includes bibliographical references and index.
ISBN 978-1-4042-4285-2 (library binding)
1. Sound—Juvenile literature. 2. Hearing—Juvenile literature.
I. Title.
QC225.5.C478 2008
534—dc22

2007032642

Manufactured in China

Contents

What is sound?

EVIDENCE

The world is filled with sound and we use sound in many ways. For example, speaking to each other is perhaps the most common form of communication, and making music is a popular form of entertainment.

Sounds can warn of danger, such as when a siren sounds from a speeding fire engine. Some natural sounds, such as birdsong, are beautiful. Not all sounds are useful, entertaining, or beautiful. Some sound is called noise and can be annoying or even damage a person's hearing. Sometimes we hear sounds that are very quiet—they may be just the sound of softly falling rain, the wind in the trees, or the sound of our own breathing.

Whatever the sound is like, all sounds travel through the air as vibrations, disturbing the air. These vibrations are picked up by your ears—and are interpreted by the brain as sounds.

When you walk down a crowded street, sound comes at you from every direction. The human brain can identify different sounds even when it hears them all together.

INVESTIGATION

Can you make a banger?

MATERIALS

A piece of smooth, glossy paper about 12 x 16 in. (30 x 40 cm) (an even larger sheet will work better).

INSTRUCTIONS

1 Fold the paper in half lengthwise, unfold it again, and then fold in the corners.
2 Fold the paper back in half again.
3 Fold in half again, so the sharp points meet.
4 Fold the top layer up, against side A. Repeat on the opposite side to create a triangular shape. Your banger is now ready to use.

1
2
3
4
A

Hold the two loose corners of your banger tightly between your thumb and first finger. Now, with a sudden movement, fling the banger down, keeping hold of the two corners. The piece folded inside the triangle should fly out at high speed, making a "bang." (It may take some practice to get it right.)

Why do you think the banger makes a loud sound?

FURTHER INVESTIGATION

What is the quietest sound you can think of? What is the loudest sound you can think of? What are the highest and lowest notes you can sing?

Some of the quietest sounds that you can hear are dry leaves rustling, a watch ticking, and wind rushing through the grass.

What makes sound?

EVIDENCE

The sound of your voice is made by two small, moist strips inside your throat called vocal cords. When you speak or sing, your vocal cords move to and fro hundreds or even thousands of times every second as you breathe out. This rapid movement to and fro is called vibration, and it disturbs the air around the vocal cords. The disturbance of the air is what sound actually is—this is why you cannot see sound.

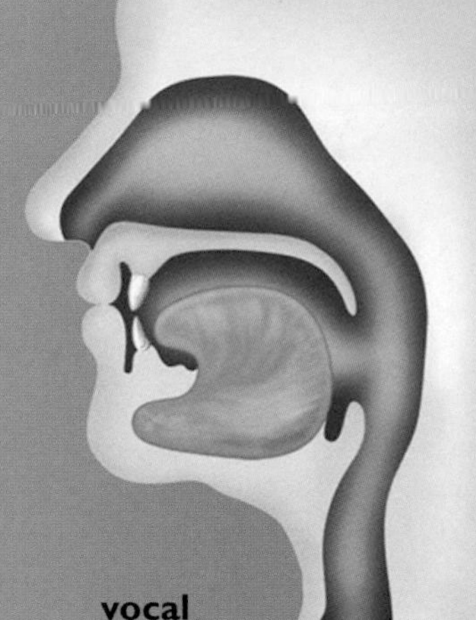

The vocal cords are part of the larynx, or voice box, in your throat.

Most sounds are caused by vibrations: a tabletop vibrates when you hit it; a stretched rubber band vibrates when you pluck it; even the air inside a flute vibrates when you blow into it. The air can be disturbed in other ways: for example, an explosion makes the air expand quickly rather than vibrate. This is called a *shock wave*. The sound of thunder is also made by a shock wave: when lightning strikes between a cloud and the ground, or between two clouds, it heats the air to a high temperature. This causes the air to expand very quickly, just like during an explosion.

Speech and other vocal sounds, such as singing, are made partly by the vocal cords, but also by the tongue, lips, teeth, and palate.

INVESTIGATION

Can you feel the vibrations?

MATERIALS

A plastic ruler, a balloon, a CD/radio with speakers, a glass bowl and a wooden spoon.

In each case, try to imagine the objects moving to and fro very quickly.

INSTRUCTIONS

Press one end of the ruler firmly down on the edge of a table, push the other end down, and then release it so the free end moves up and then down again. Can you see the vibrations?

Gently tap the side of the glass bowl with the wooden spoon, causing the bowl to vibrate and so "ring" for a few seconds. You can stop the vibrations—and the sound—by gently touching the bowl.

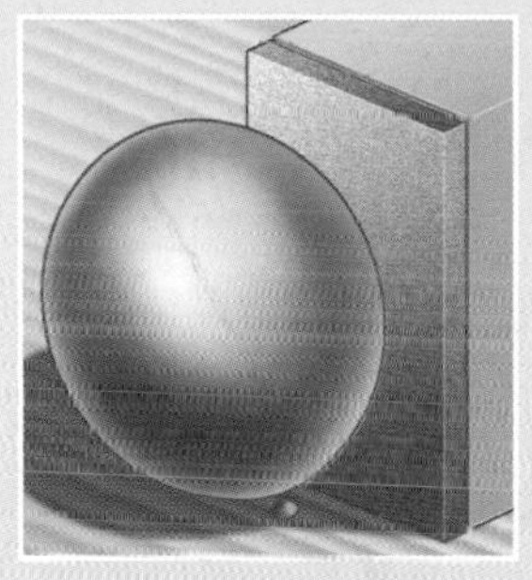

Blow up the balloon and tie its end. Now play some music on the CD/radio—not too loudly—and then hold the balloon lightly against the speaker and place your lips against the balloon. The speaker vibrates to produce sound, and those vibrations make the balloon vibrate, too. Can you feel the vibrations?

FURTHER INVESTIGATION

Why do you think you can see the ruler vibrating but not the side of the glass bowl? Does the glass bowl vibrate more quickly or more slowly than the ruler? Can you work out what it is that is vibrating inside the speaker?

The explosion caused by the demolition of this tower creates a shock wave. The shock wave disturbs the air and when it reaches your ears, you hear a loud bang.

How does sound travel?

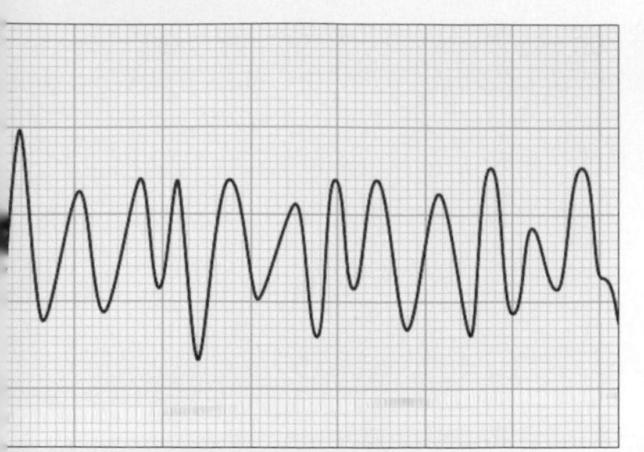

If you could see the sound waves of a bell, they would look a little like this. The pattern of high- and low-pressure air is complicated. This is because a bell vibrates in a way that depends on the shape and size of the bell itself.

EVIDENCE

The sides of a ringing bell move outward and inward hundreds of times every second as the bell vibrates. Each time the side of the bell moves outward, it "squashes" the air next to it. As it moves back in again, the air "stretches" to fill the space it leaves behind it. The air in a party balloon is "squashed," and squashed air is said to be under high pressure. The air in a vacuum cleaner is "stretched" by a fan inside the machine, and is said to be under low pressure.

Sound waves are waves of changing air pressure. They travel very fast—at more than 985 ft. (300 m) per second. This means that the speed of sound is more than 12 times the speed of a car traveling at 62 mi. (100 km) per hour. Sound does not only travel through the air: in fact, sound travels much faster through liquids and solids than through gases such as the air.

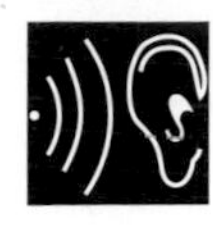

INVESTIGATION

What does a sound wave look like?

MATERIALS

A large bowl, water, and a flashlight.

INSTRUCTIONS

Sound waves are invisible, but you can see how they work by making ripples in a bathtub or a large bowl of water. Fill the bowl with water, and wait until the surface of the water is flat.

Shine the flashlight down on the surface of the water, so that you can see the flashlight light at the bottom of the bowl. Gently put a finger into the water, near the middle. Move it slowly to and fro. Can you see the ripples?

Just as your finger disturbs the surface of the water, sources of sound disturb the air. The disturbance travels out in all directions, just as the ripple does in water. Did you notice how the ripples in the bowl of water bounced off of the sides of the bowl? Sound waves can bounce off of surfaces, too. Can you hear sound that has bounced?

Dolphins make moaning, screeching, and clicking sounds underwater to communicate with each other. The sound they make travels about 1 mile (1.5 kilometers) per second, and can be heard by other dolphins hundreds or even thousands of miles away. Underwater noise made by humans can confuse and even harm dolphins.

FURTHER INVESTIGATION

In a thunderstorm, why do you hear the thunder a few seconds after the lightning?

Does sound travel through solid materials? Try pressing your ear against a tabletop and gently tapping the table with a pencil.

Can sound bounce?

EVIDENCE

When sound waves hit a solid surface, such as a wall, they bounce back again, or "reflect," just like the water waves in the activity on page 9. If you shout loudly at a wall that is more than about 55 yards (50 m) away, you should hear the reflected sound as an echo of the sound of your voice. From a wall 55 yards (50 m) away, it arrives back at your ears about one third of a second after you shouted. If the wall is much closer than this, the reflected sound arrives back at your ears after just a tiny fraction of a second, so that you hear the original sound and its echo together.

Reflected sound or echoes occur where sound bounces off of walls and other solid surfaces. When sounds in a small space echo and re-echo many times, all the echoes mix together, so that sounds just seem to "hang" in the air. This is called *reverberation*, and you will recognize it if you've ever shouted into the opening of a cave.

The hard walls, floor, and ceiling mean that sounds reflect many times inside this building. This is why even a softly spoken word in such a place reverberates, as if your voice "hangs" in the air.

INVESTIGATION

How well can you hear reflected sound?

MATERIALS

A hardback book, a folded towel, and your home.

INSTRUCTIONS

Hold a hardback book about 4 in. (10 cm) in front of your face and speak or sing. Close your eyes and concentrate on the sound. Take the book away from your face and speak or sing again. How does it sound different?

Now hold a folded towel 4 in. (10 cm) in front of your face, and speak or sing again. It should sound different, because soft materials such as towels do not reflect sound as well as hard objects, such as the book.

Go outside, stand somewhere away from walls, and clap your hands. There are no solid surfaces, so you should hear no reflected sound.

Now, go inside and clap in different rooms. Where does the sound of your clapping "hang" in the air?

Where does the sound reflect better—in the rooms with soft furnishings, or solid surfaces, such as walls and hard floors?

FURTHER INVESTIGATION

Repeat the investigation using other objects. Try things made of different materials, such as a metal saucepan. Try large objects and small objects.

Where else might you hear reflected sound?

Bats send out pulses of sound that bounce off of objects nearby. By listening for echoes, they can find their way around in the dark and even hunt for insects.

How do we hear?

EVIDENCE

As sound spreads out in all directions from a sound source, your ears may pick it up. Some sound enters and passes down a tube called the *ear canal.* At the end of your ear canal is a thin piece of taut skin called the *eardrum.* The sound waves cause the eardrum to vibrate, and these vibrations pass on to three tiny bones—the smallest bones in your body. These bones carry the vibration to the *cochlea*, an organ that looks like a snail shell but is soft. The cochlea is filled with fluid and is lined with thousands of tiny hairs. The three bones are in contact with the cochlea and they make the fluid vibrate; this makes the hairs vibrate, too. The hairs are connected to nerves, which carry messages about the vibrations to the brain. This is how we hear sound.

The shape of the outer ear helps to channel sound waves into the ear canal.

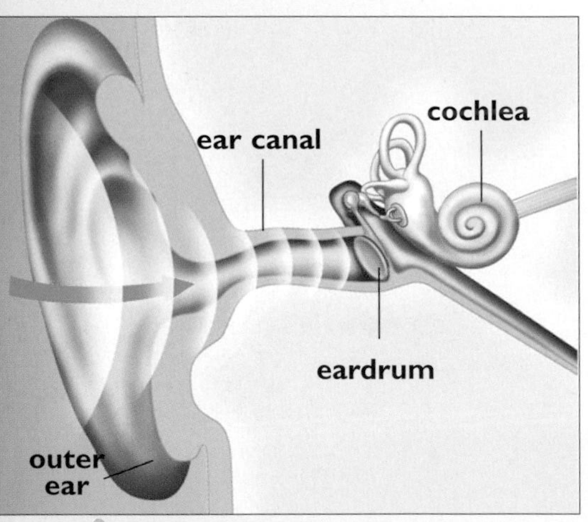

Once inside, the sound waves pass down the ear canal to the eardrum, causing it to vibrate.

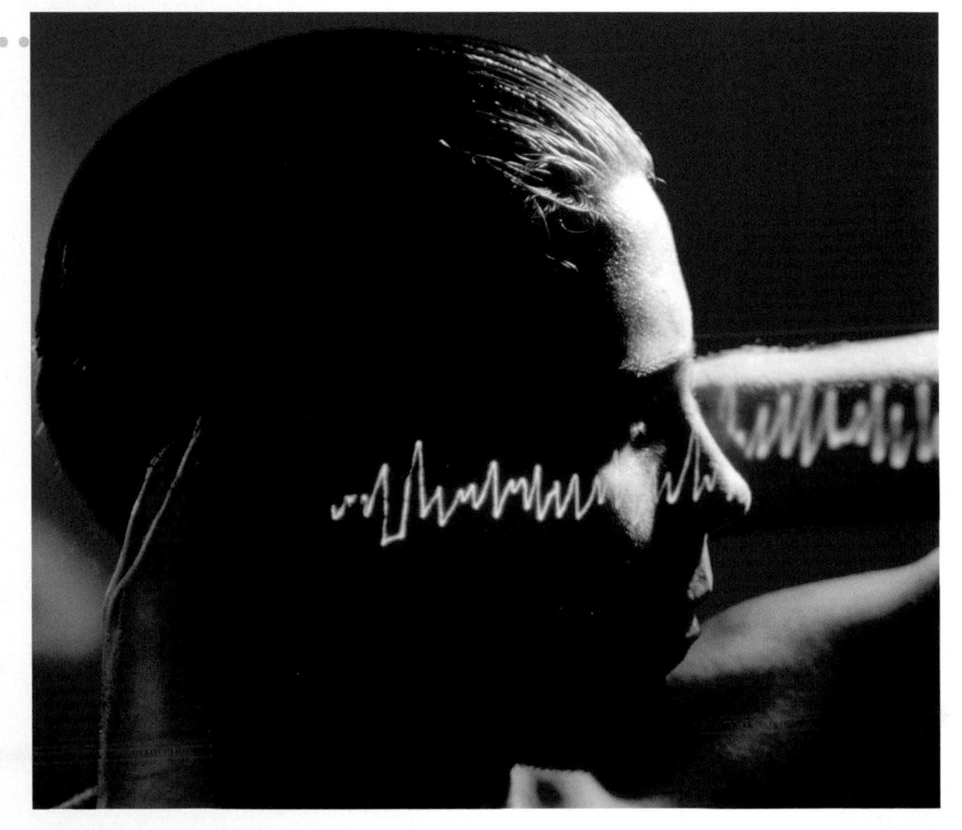

Most animals can hear. Some have ears similar to our ears, but others, such as grasshoppers, hear in different ways.

INVESTIGATION

How do sound waves make the eardrum vibrate?

You can see for yourself how sound waves can make a stretched surface vibrate, by making a model of an eardrum.

MATERIALS

A balloon, a balloon pump, scissors, a small glass bowl, uncooked rice, and tape.

INSTRUCTIONS

Blow up the balloon to stretch the rubber, and then let the balloon deflate. Cut a piece of rubber from the balloon that is bigger than the opening of the bowl.

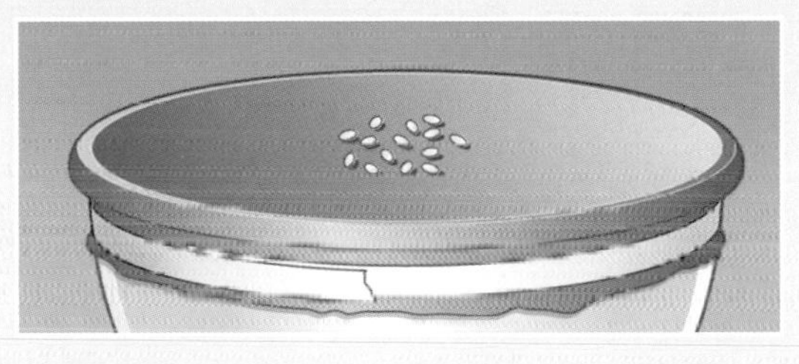

Tape the rubber around the top of the bowl so that it is stretched quite taut over the top, like a drum skin.

Place about ten rice grains on the rubber and shout loudly at the rubber. Try not to blow the rice away as you shout.

Some of the sound waves you produce when you shout hit the rubber and make it vibrate. What happens to the rice grains?

A grasshopper has a piece of stretched skin—like an eardrum—on its abdomen (rear body part) or in some species, on its legs. The stretched skin of the "ear" detects sound waves.

FURTHER INVESTIGATION

Try making the rubber sheet "hear" other sounds: try speaking more quietly or shouting really loud, or place the bowl next to one of the speakers of a CD/radio.

Why are some people hard-of-hearing?

EVIDENCE

You may be deaf or you may know someone who is deaf. Some deaf people have been unable to hear since they were born, but others have become deaf during their lives. There are several causes of deafness. In some people, the nerves that lead from the cochlea (see page 12) to the brain may be damaged, so that sound signals do not reach the brain. In other people, one of the three tiny bones may be stuck, so that little or no sound reaches the cochlea. Even earwax can cause partial deafness, by blocking the ear canal. People gradually lose their ability to hear as they grow older.

If you are deaf, perhaps you could ask a hearing person to try the investigations in this book with you, so that they can describe to you what happens in each one—you don't have to be able to hear to understand how sound works.

People who spend a lot of time in noisy places must protect their ears from damage. They often wear ear defenders, which look like large headphones, to prevent most of the sound from getting to their ears.

INVESTIGATION

What materials would you use to make ear defenders to wear in a noisy environment?

MATERIALS

A portable radio, a thick woolen sweater, and a selection of other materials, such as: a cardboard box, styrofoam packing, a large wooden board, and a sheet of kitchen foil.

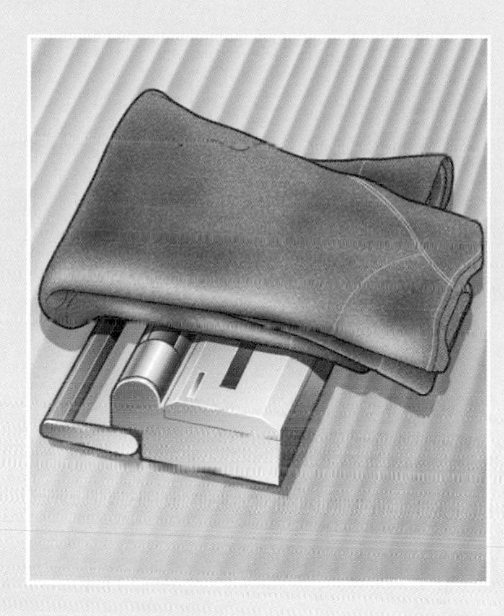

INSTRUCTIONS

Turn on the radio, making sure that it is not too loud.

Hold the woolen sweater against the radio's loudspeaker. Does this make any difference to the level of sound?

Fold the sweater in two or three, and repeat the second step. Does the sound become even more quiet?

Now try placing the other materials over the radio's loudspeaker—try each one on its own and try different combinations. What works best at reducing the sound level?

Deaf people can communicate just as well as hearing people. Most use sign language, moving their hands and making facial gestures to represent words.

FURTHER INVESTIGATION

Sound can escape through gaps. Compare the level of sound you can hear from a room with the door completely closed to the same room with the door slightly open.

What materials might you use to soundproof a room? Sit still in a quiet room for a few minutes. You should find that you begin to hear even the quietest sounds, as your brain adjusts to the quiet.

Why are some sounds louder than others?

EVIDENCE

Things that disturb the air produce sounds; the greater the disturbance, the louder the sound. Hitting a drum hard makes a louder sound than tapping it gently, because the drum skin vibrates more vigorously the harder you hit it. This disturbs the air more than a gentle vibration would. The drum would not sound so loud if it did not have the wooden body. This is because the drum skin passes on its vibrations to the body, and that, too, disturbs the air.

The body of a guitar acts in the same way, making the sounds made by the strings much louder. Because loud sounds disturb the air more than quiet sounds, they can cause damage to your ears. The hairs in the cochlea inside the ear become damaged, and that can lead to a gradual loss of hearing. So, loud sounds can be dangerous, but they can also be simply annoying. For both reasons, it is important to monitor sound levels in noisy environments, such as in factories.

Rocket engines are very loud, because explosions happen inside them. But even a loud sound will sound quiet if you are far away.

INVESTIGATION

How can you make a heartbeat sound louder?

MATERIALS

A plastic funnel, a clean plastic tube about 20 in. (50 cm) long (part of a hosepipe will do), and tape.

INSTRUCTIONS

Push the small end of the funnel into the tube, and connect the two objects with tape. Gently put the other end of the tube to your ear.

Now, place the open end of the funnel on your chest, over your heart (slightly to the left as you look down at your chest).

You should be able to hear your heartbeat as a low thud. This is because the sound passes into the funnel, and bounces along inside the tube, without spreading in all directions.

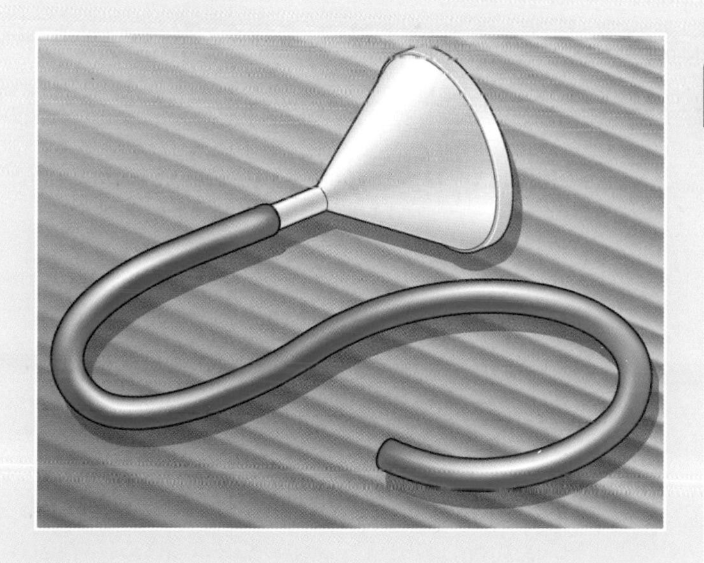

A megaphone is used for amplifying sounds. A megaphone is shaped in this way so the sound is focused in one direction, which makes it carry farther away.

FURTHER INVESTIGATION

You can make a megaphone by making a cone shape from a piece of card.

See if you can make the prongs of a fork ring by tapping the fork on a hard surface. What happens if you hold the fork, prongs up, against a tabletop? Why does the ringing get louder?

See how much louder very quiet things sound when you are very close to them.

Why are some sounds more high pitched than others?

Women's vocal cords can vibrate at higher frequencies than men's, because they are smaller. During teenage years, men's vocal cords grow much longer, but women's stay about the same length. The Adam's apple is the result of this growth.

EVIDENCE

Most sources of sound—such as a bell or your vocal cords—produce sound because they vibrate. The faster they vibrate, the more high pitched the sound they produce. For example, the reason why most women can sing higher notes than most men is that women's vocal cords can vibrate more quickly than men's. The number of times an object vibrates every second is called the *frequency of the vibration*.

The most high-pitched sounds humans can hear are produced by things that vibrate at about 20,000 times every second. Many things vibrate at higher frequencies than this. We do not hear the sound they make, but many animals can. This kind of sound is called *ultrasound*. The very lowest sounds humans can hear are produced by objects vibrating about 20 times every second.

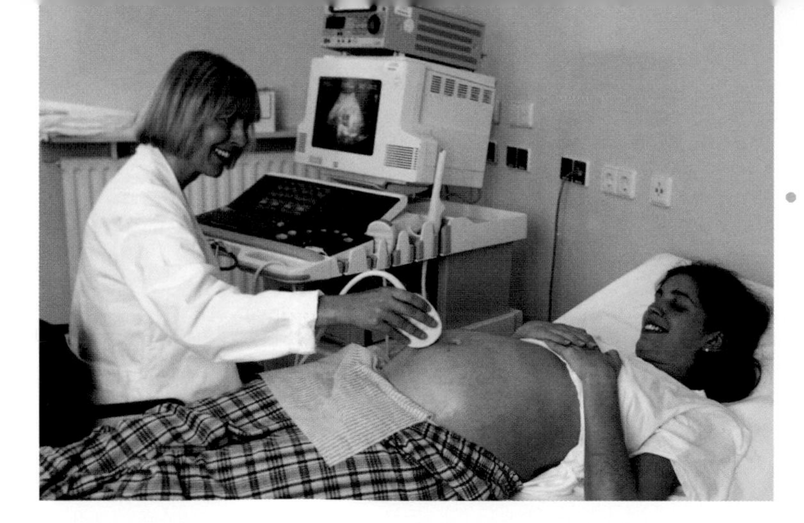

This ultrasound scanning machine has a probe that produces ultrasound. The ultrasound travels through to the womb where there is a growing baby. Some of the sound reflects off of the baby, and the machine detects the reflected sound and produces a picture of the baby.

INVESTIGATION

How do the treble and bass controls change the low and high frequencies of music?

MATERIALS

A CD/radio (FM) with treble and bass controls. Ask an adult before using the CD/radio, and if you need help.

INSTRUCTIONS

Turn on the CD/radio so that you can hear an FM radio station. Now tune the radio to a point in between stations, so that you hear a hissing sound. This sound is called *white noise,* and it contains low, medium, and high frequencies of sound.

Turn the treble control all the way up—this boosts the high frequencies. How does the sound change?

Turn the treble control all the way down, to get rid of most of the very high frequencies. Now how does it sound? Repeat the last steps, but this time with the bass control.

How do you think the treble and bass controls might affect the sound of a piece of music? Try it, by putting on a CD.

FURTHER INVESTIGATION

Try listening to some rock music as you adjust the treble and bass controls. If there are cymbals in the music, do you hear them best when the treble control is high or low?

Do all frequencies of sound pass equally well through different materials? Try listening to some music that is playing in the next room—do low or high frequencies pass through walls better?

What makes things sound different from each other?

EVIDENCE

If someone plays a note on a piano, you can tell that it is being played on a piano and not being produced by a bell or a gong. What is it that gives different sounds their character, their identity? It has to do with the way things vibrate. The air in a flute vibrates very smoothly and simply, but the strings inside a piano vibrate in a much more complicated way, so a piano and a flute sound very different from each other. In fact, each note played on a piano is produced by several strings vibrating at the same time, making each note a mixture of several different frequencies of vibration.

Most sounds are complicated mixtures of different frequencies, which mix together to make a distinctive sound. Some sounds are produced by mixtures of so many different vibrations that they do not sound like pure notes at all. Bangs, clicks, and hissing sounds are good examples of this.

Most instruments in the orchestra can make sounds over a range of pitches. Large instruments, such as cellos, bassoons, and tubas, make a range of low-pitched sounds, and smaller instruments, such as violins, piccolos, and trumpets, have higher-pitched ranges.

INVESTIGATION

Sounds and surfaces

MATERIALS

Thick paper or thin card, a sewing needle, and tape.

INSTRUCTIONS

Make a cone out of the paper or card, holding it together with tape. Stick the needle about halfway through the end of the cone, and secure it with another piece of tape. Then hold the cone to your ear as you scrape the needle across different surfaces: for example, a newspaper, sandpaper, brick, and pebbles.

What do you hear? Can you relate the sound to the surfaces? Is it equally easy to scrape the needle across all the surfaces? What happens if you move the needle faster?

A *spectrum analyzer* is a device that breaks down complicated sounds into a mix of simple sounds, each with its own frequency. In this picture, the sound of a bell is being analyzed in this way.

FURTHER INVESTIGATION

If you and your friends have some musical instruments, try to find a single note that you can all play together at once. Or you could find a note that you can hum, whistle, and sing. Since these notes are all in tune, why do they sound different?

How do stringed instruments produce sound?

EVIDENCE

You may have noticed that some musical instruments have strings. Any stretched string vibrates when it is plucked, strummed, hit, or bowed. The note produced by a vibrating string depends upon three things: how tightly the string is stretched, how heavy the string is, and how long it is. To make a string produce a very high-pitched note, you would make sure that it was stretched tightly, that it was very light, and that it was very short. Long, heavy, loose strings produce low notes.

Although the strings of a guitar are all the same length, they can be made shorter or longer by pressing them against the fretboard. The farther up the fingerboard you press the string, the shorter the length of the string that can vibrate and the higher the note.

On a guitar, all the strings are the same length, but the strings that produce the lower notes are much thicker than the ones that produce high notes. The strings can be tightened or loosened so that they can be tuned to make just the right notes. A vibrating string does not produce a very loud sound, so most stringed instruments have a large, empty wooden body that vibrates with the string, disturbing more air and making the sound much louder.

INVESTIGATION

Can you make a stringed instrument?

MATERIALS

A shoebox, four large rubber bands, two pencils, and some tape.

INSTRUCTIONS

Stretch a rubber band between two fingers and pluck it with your other hand. You should hear a note. Change how tightly you are stretching the rubber band, and see how the note changes.

To make the guitar, cut a hole about 2 in. (5 cm) wide in the middle of the lid of the shoebox. Tape one pencil at each end of the lid. Put the lid on the box.

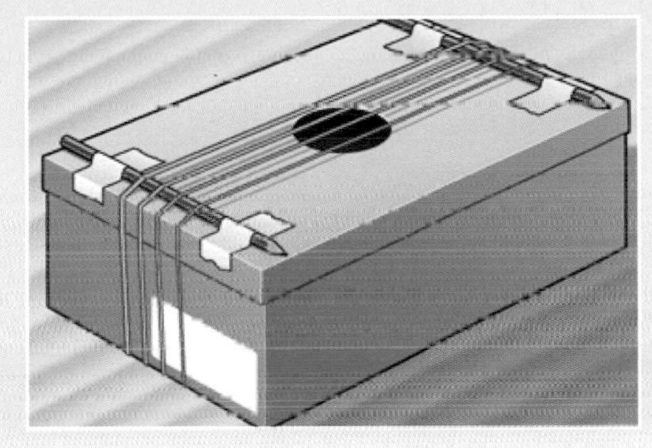

Stretch the rubber bands over the pencils and around the shoebox, and pluck each band, one at a time.

You can make higher-pitched notes by increasing the tension of the rubber bands or decreasing their length. Pressing a band down to the box's surface (just like pressing a real guitar string on a fretboard) will reduce the length, and pulling the band a little way away from the side of the box will increase the tension.

FURTHER INVESTIGATION

In the investigation, try thicker rubber bands—does that make the note higher or lower? What difference does the box make to the sound of the rubber bands?

Unlike a guitar, a piano has no fretboard. Its strings are all different lengths and thicknesses, so each string vibrates at a different frequency. Hitting a piano key makes a hammer hit a certain group of strings, which vibrate and produce a note.

How do wind instruments produce sound?

EVIDENCE

Some musical instruments produce sound because the air inside them vibrates. In a flute or recorder, simply blowing in the right way disturbs the air to produce the sound. In a saxophone, you have to blow past a sliver of wood called a *reed*, which vibrates and disturbs the air. A trumpet player vibrates his or her lips against the mouthpiece, so the air in the tube of the instrument vibrates.

The note made by a wind instrument, such as a recorder or a flute, depends upon the length of the instrument. You can change the length of the vibrating column of air inside a recorder by covering up different holes along its body. This is how a recorder player changes the note that is being played. A trumpet works in a different way: blowing harder or softer through the lips causes them to vibrate at different frequencies, producing different notes. There are also three keys that change the length of the tube by closing off different sections of it.

Each of the pipes of this church organ is a different length. The longest ones produce very low-pitched notes, and the shorter ones produce higher-pitched notes. Because the pipes also have large diameters, they disturb a lot of air, making the sound they produce very loud.

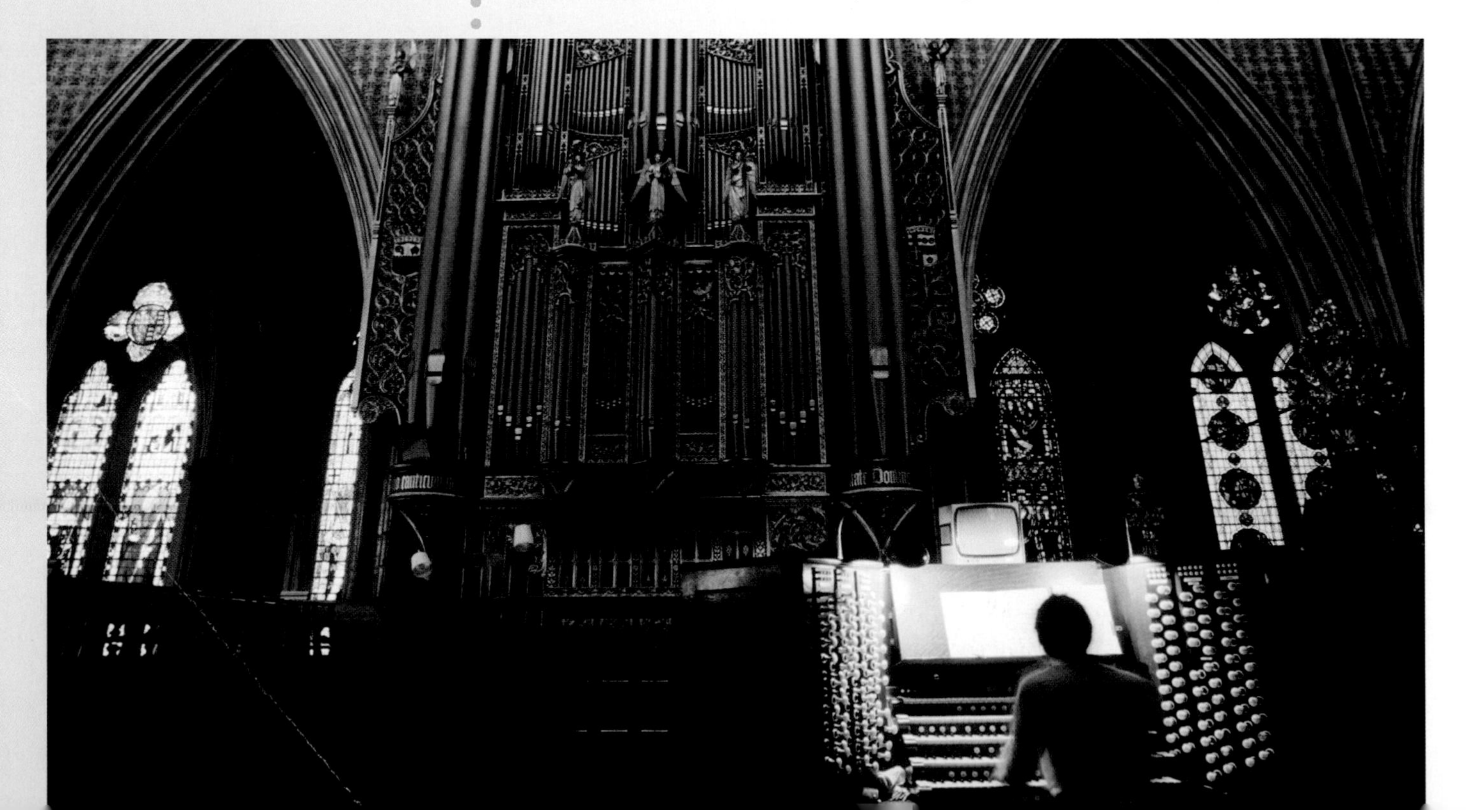

INVESTIGATION

How does the length of the column of air affect the note?

MATERIALS

A piece of garden hose (or tube 1/2–1 in. or 1–3 cm wide) about 12 in. (30 cm) long.

INSTRUCTIONS

Hold the hose vertically in the water, so that about 1/2 in. (1 cm) of it is under water. Hold your mouth next to the top of the hose and blow across the top. You should hear a sound. Try lowering the hose into the water a little. It now contains a shorter column of air, which should give a higher note. The air in the hose vibrates when you blow across the top. The frequency at which it vibrates depends upon the length of the column of air.

FURTHER INVESTIGATION

Go outside with a 1 yard piece of hose. Make sure no one is standing close to you. Now whirl the hose around your head. What do you hear?

Pour water into a bottle and fill it almost to the top. If you blow across the top you should hear a note. What happens if you empty some of the water and try again? Can you arrange several bottles, with different amounts of water in them, to make up a tune you can recognize?

The pipes of this panpipe are all of different lengths. Blowing across the top produces a different note in each pipe.

How do percussion instruments produce sound?

EVIDENCE

Percussion instruments are instruments that you hit, rub, or rattle to make a sound. They are usually used to help keep the rhythm of a piece of music. Most percussion instruments have no particular pitch—they do not produce musical notes like wind and string instruments do. Banging a drum or crashing a cymbal produces many different frequencies of vibration, making banging or crashing sounds instead of musical notes.

Music at a carnival. The steel drums are tuned to particular notes.

There are some percussion instruments that are tuned to produce musical notes—one example is a xylophone. Even some drums are tuned to particular notes—you may have heard of timpani and steel drums. Most drums have a stretched skin that vibrates when it is hit with a hand or a stick. Other percussion instruments, such as cymbals and cowbells, do not have stretched skins. Instead, they are made of a material, such as metal, which vibrates when you hit, rub, or rattle it.

INVESTIGATION

Can you make chimes?

MATERIALS

Different-sized items of old metal cutlery, such as spoons, forks, and knives, string, tape, and scissors.

INSTRUCTIONS

Cut five pieces of string, each about 4 in. (10 cm) long, and attach one end of each piece firmly to a different piece of cutlery with tape.

Hang the cutlery by sticking the free ends of the string from the bottom of a kitchen wall cupboard or the edge of a table.

Now hit each piece of cutlery with another metal object (another spoon will do). Does each item sound the same when you hit it? Which ones ring for the longest time? Try hitting the cutlery chimes with other materials, such as wood.

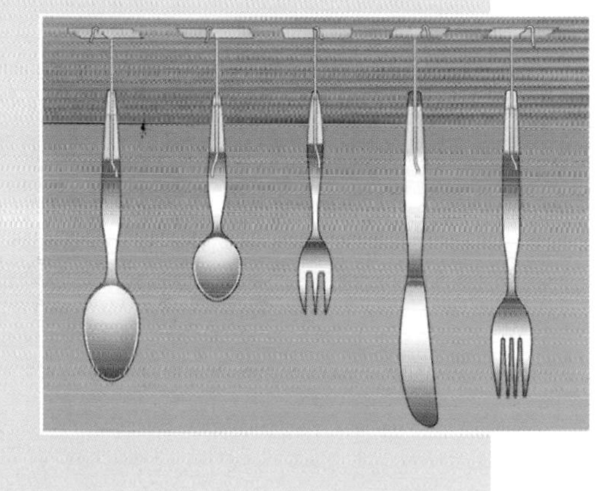

Cymbals are slightly curved disks that are struck together to make a ringing sound.

FURTHER INVESTIGATION

Try making some percussion instruments of your own design. You could make shakers by filling plastic bottles with rice or lentils. Think about what is vibrating when you play the instruments.

How can you make the sound of your percussion instruments louder?

Make your cutlery chimes into wind chimes, by hanging them very close together outside. When the wind blows, the chimes should play.

How do microphones work?

EVIDENCE

Microphones are used in recording studios. The signals from microphones next to different instruments and voices can be mixed together to produce the recording.

You may have an iPod, an MP3 player, or CDs that allow you to play back recorded music or speech, or perhaps you use a computer to download music from the Internet. To record sound, you have to be able to detect sound waves, and that is usually done with a microphone.

A microphone is an electric "ear." A coil of wire is attached to a diaphragm and positioned inside a magnet. When sound hits the diaphragm, it vibrates. The vibration of the coil in the magnet produces electrical signals in the wire.

There are several different types of microphone, but they almost all have a thin sheet of material called a *diaphragm*, which vibrates like your eardrum does when sound hits it. The vibration of the diaphragm causes an electric current to flow backward and forward through the wires that are connected to the microphone—the electric current is an exact copy of the variation in pressure in the sound waves. The electric current is called a *signal*, and can be recorded in many ways—for example, as patterns of holes on a CD, or as a series of numbers in a computer—or it can be made stronger and sent to loudspeakers.

INVESTIGATION

How does a microphone pick up sounds?

MATERIALS

A music system with an external microphone, or a cassette tape recorder or dictaphone with a built-in microphone.

INSTRUCTIONS

Record the sound of your voice and other sounds and play them back. Try recording some sounds close to the microphone and some far away.

When you play back the sounds you have recorded, adjust the treble and bass controls if you can, as in the investigation on page 19.

Try covering the microphone with various materials—try a woolen scarf and a small cardboard box, for example. How does the sound of the recording change?

The signal from a microphone is connected to an amplifier, which makes a powerful copy of the electric signal. This can be sent to loudspeakers, where it makes large paper cones vibrate—in exactly the same way as the original sound.

FURTHER INVESTIGATION

Use reference books to find out about the different types of microphone and how they work, and how sound signals can be held on CDs and tapes, and in computers.

Why do things sound quieter when the microphone is held farther away? Think back to the ripples in the bowl of water, and how they spread out.

Glossary

Air pressure
The force exerted by the air over an area. Compressed (squashed) air has a high pressure, because it exerts a large force on whatever container it fills. The air pressure varies along a sound wave.

Cochlea
A spiral-shaped organ, about the size of a small snail, inside the ear. Vibrations picked up by the eardrum are passed to the cochlea. Hairs inside the cochlea are connected to nerves that send signals to the brain.

Decibel (dB)
The unit of the power of sound. A sound of 0 dB is only just audible, and a sound of 120 dB is painfully loud.

Ear defenders
A device that looks like a large pair of headphones, which protects ears from dangerously loud sounds.

Eardrum
A thin piece of skin inside the ear canal that vibrates when sound hits it. Its vibrations are passed on to the cochlea.

Echo
A sound that has bounced off of a surface.

Frequency
The number of vibrations of an object or a sound wave each second. A high frequency sound is high pitched.

Loudspeaker
A device that produces sound when electrical signals pass through it.

Microphone
A device that is used to convert a sound wave into an electrical signal. The changes in the signal are the same as the changes in air pressure in the sound wave.

Reflection
The throwing back of a wave, such as a sound wave, by a surface. Echoes are produced by reflection.

Reverberation
A word that describes the way sound seems to "hang" in the air in large halls or caves.

Shock wave
A sudden change in air pressure, caused by a large disturbance of air. A powerful shock wave creates a loud sound.

Sound wave
A series of very rapid changes of air pressure, which travels through a solid, liquid, or gas. We hear sound when the sound wave meets our ear.

Spectrum analyzer
A device that breaks down a sound wave into the sound frequencies that make it up, and measures the strength of each frequency.

Speed
The rate at which something travels. Scientists usually measure speed in meters per second (m/s). Sound travels at a speed of about 330 m/s in air.

Ultrasound
Sound with a frequency higher than the human ear can hear.

Vibration
A rapid movement to and fro. Vibrations often produce sound waves.

Vocal cords
A pair of flaps of skin inside your windpipe. The vocal cords vibrate as air rushes past them, making the sound of your voice.

Further information

BOOKS

DK Science Encyclopedia
(DK Children, 1999)

Light and Sound (Science Files)
by Chris Oxlade
(Hodder Wayland, 2005)

Light and Sound (The Young Oxford Library of Science)
by Jonathan Allday
(Oxford University Press, USA, 2003)

Rubber-Band Banjos and a Java Jive Bass: Projects and Activities on the Science of Music and Sound
by Alex Sabbeth
(Jossey-Bass, 1997)

Sound (Science Answers)
by Chris Cooper
(Heinemann Library, 2004)

Sound (Science Around Us)
by Sally Hewitt
(Chrysalis Education, 2003)

Sound (Science Files)
by Steve Parker
(Heinemann Library, 2005)

Sounds Dreadful (Horrible Science)
by Nick Arnold
(Scholastic Hippo, 1998)

CD-ROMS

Eyewitness Encyclopedia of Science
Global Software Publishing

I Love Science!
Global Software Publishing

ANSWERS

page 5 When you fling your banger down, the piece that is folded inside the triangle flies out and disturbs the air, making a loud bang.

page 7 The side of the glass bowl vibrates faster than the ruler—too fast to see.
The cone inside the speaker is vibrating.

page 9 A sound that has bounced is an echo.
The thunder is produced at the same time as the lightning, but the light arrives at your eyes before the sound arrives at your ears.
The tapping is much louder when your ear is pressed to the table, showing how well sound travels through solids.

page 11 The clapping sounds different in different rooms. In rooms with soft furnishings—such as beds, carpets, cushions, and heavy curtains—the sound does not reflect very well. Sound reflects much better off solid surfaces, such as walls and hard floors.
You can hear reflected sound—echoes—wherever there is a large solid surface. For example, inside a cave, in a church, in a large, empty bathroom.

page 13 You can see that the rubber is vibrating, because the rice grains jump up and down.

page 15 The best materials to use for blocking sound are heavy materials that also contain pockets of air. The sound bounces around inside the air pockets, losing a bit of energy with each bounce. Real ear defenders are made from a heavy foam material (with air pockets) and a thick layer of plastic.
Rooms such as recording studios must be soundproofed. They often have very thick walls, or double walls with foam between them. The doors are heavy and double-glazed.

page 17 The vibrations of the fork pass to the tabletop, and make the whole surface vibrate. This disturbs much more air, making the sound much louder.

page 19 The sound made by cymbals is mostly high frequencies. The sound of the cymbals should be clearest when you have the treble control set high.
Low frequencies—such as the sound of a booming bass drum—pass through walls better than high frequencies.

page 21 Rough surfaces are louder than smooth ones, because the needle drags and jumps across the tiny ups and downs of a rough surface. This causes the needle to wobble, and the wobbles are converted to shuddering movements of the paper cone. These movements make the air in the cone shudder, too—which is what sound is.
You have to push harder to move the needle across a rougher surface, because some of your push turns into sound energy.
Moving the needle faster makes the wobbles (in needle, cone, and air) more frequent. Higher frequency means higher pitch, so the sounds you hear get higher the faster you move the needle.
Almost all sounds are made of many simpler sounds. Often—as in most musical instruments—one of those simpler sounds is a lot louder than the rest, and that is what gives the sound its pitch. But the other sounds present all work together to produce the distinctive quality we recognize.

page 23 Thicker rubber bands should vibrate at lower frequencies, making lower-pitched sounds.
The box picks up vibrations of the rubber bands, and it, too, vibrates, making them much louder.

page 25 Air rushes through the hosepipe and vibrates inside. You should hear a note, which should become louder the harder you whirl the hosepipe.
With less water in the bottle, a lower note is produced.

page 27 You can make the sound of your percussion instruments louder by hitting or shaking them harder, or by using harder or heavier objects—pasta instead of rice, for example.

page 29 The recorded sound will be quieter, especially the higher notes.
The microphone picks up less sound when it is farther away, because sound spreads out in all directions, like ripples on a lake.

Index

Web Sites
Due to the changing nature of Internet links, PowerKids Press has developed an online list of Web sites related to the subject of this book. This site is regularly updated. Please use this link to access this list:
www.powerkidslinks.com/sci/sound